Scholastic Success With All About Numbers

Kindergarten

by Janie Schmidt

SCHOLASTIC
PROFESSIONAL BOOKS

New York • Toronto • London • Auckland • Sydney •
Mexico City • New Delhi • Hong Kong • Buenos Aires

Cover art by Victoria Raymond
Cover design by Maria Lilja
Interior illustrations by Carol Tiernon
Interior design by Quack & Company

ISBN 0-439-44499-3

1 2 3 4 5 6 7 8 9 10 40 09 08 07 06 05 04 03

Introduction

Parents and teachers alike will find this book to be a valuable learning tool. Children will enjoy completing a wide variety of math activities that are both engaging and educational. The activities include dot-to-dots, mazes, hidden pictures, matching, and other fun ways to learn about numbers. While engaged in these activities, children are afforded practice in recognizing and counting numbers from 1 to 20. Children will also learn about one-to-one correspondence, writing number words from one to ten, and counting to 30. Take a look at the Table of Contents and you will feel rewarded providing such a valuable resource for your children. Remember to praise children for their efforts and successes!

Table of Contents

One Old Octopus

Trace and write.

1

Color each shape with 1 fish.

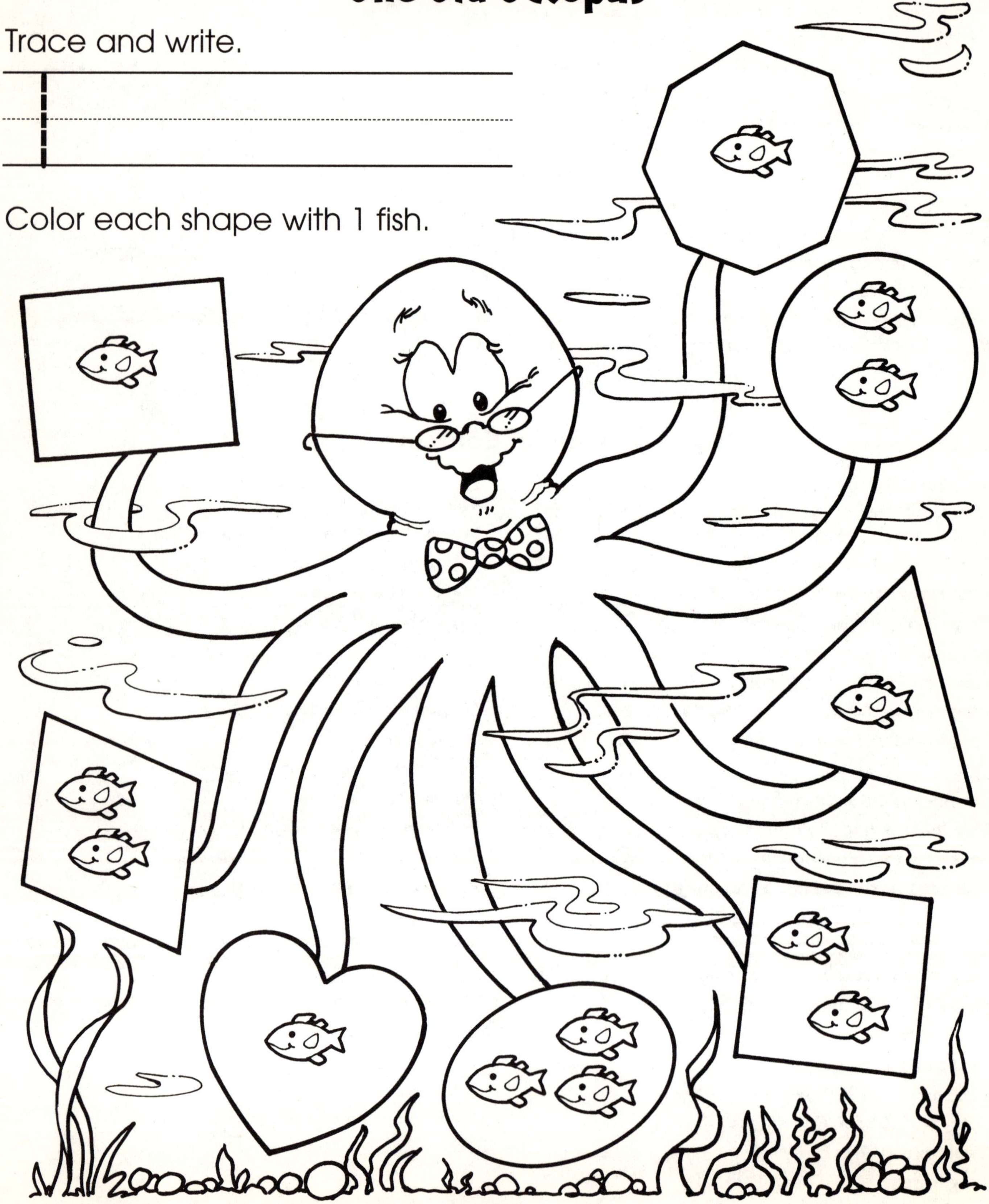

An Underwater Home

Count each group of things found in the sea. Color one of each.

How many objects did you color?

Two Talking Turtles

Trace and write.

2

Color each circle with 2 dots.

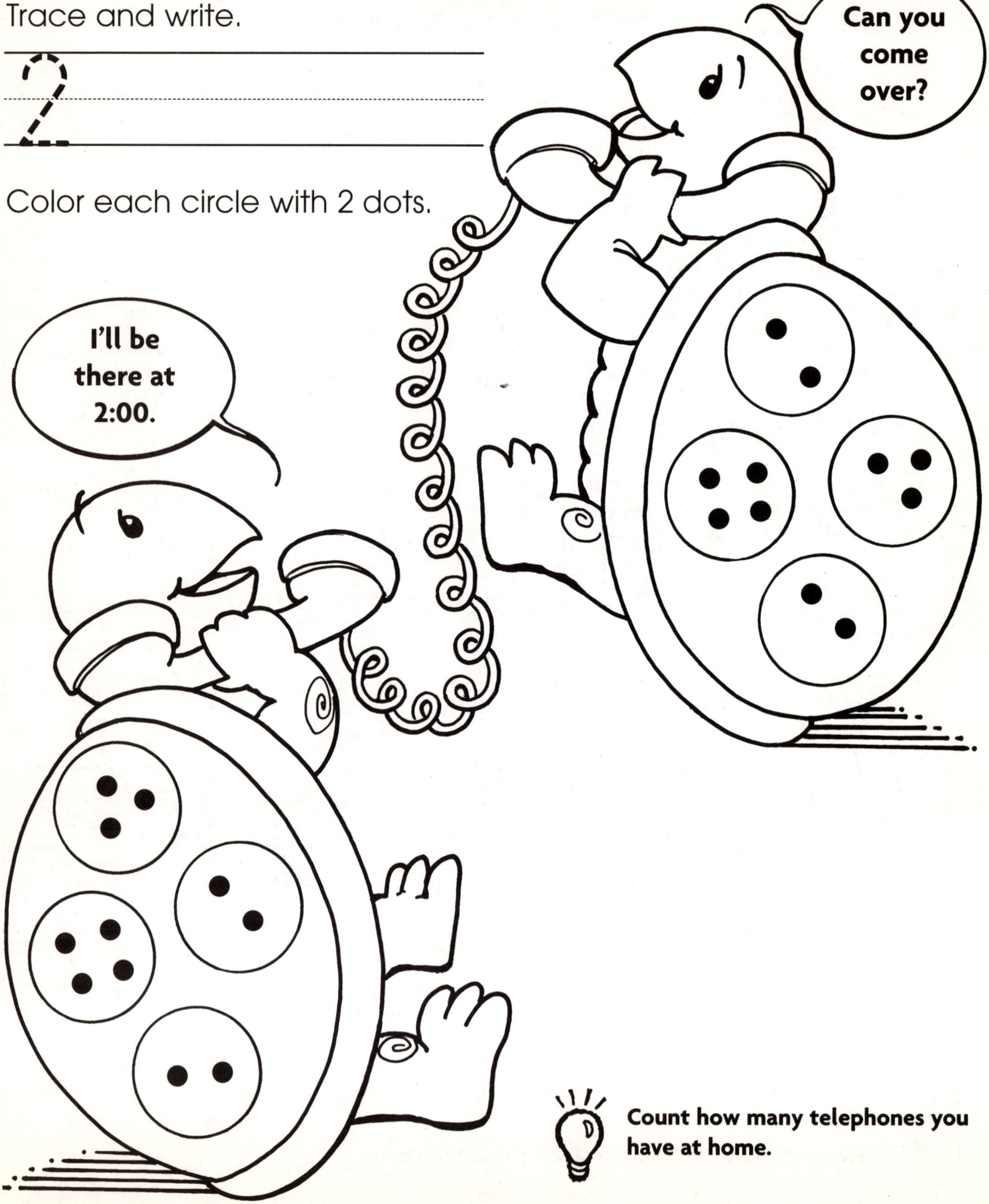

Count how many telephones you have at home.

Mrs. Tacky Turtle

Circle the number that tells how many.

	1	2
	1	2
	1	2
	1	2
	1	2
	1	2
	1	2

What else could Mrs. Turtle wear? Draw 2 of them.

Name ________________________________ **Identifying 3**

Three Tiny Tugboats

Trace and write.

Color each barge with 3 objects.

Draw 3 logs on each barge.

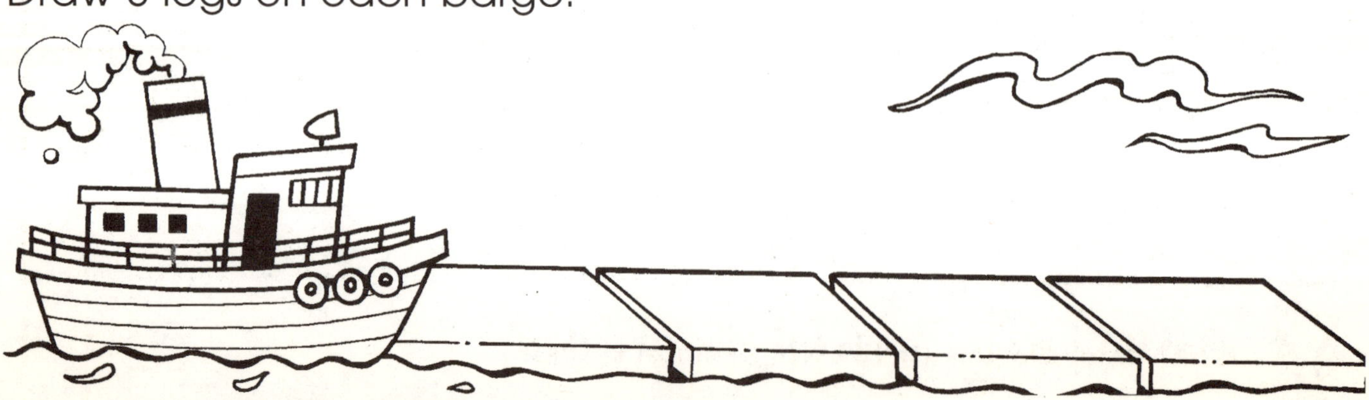

Name ________________________________

Tugboat Tow

Use the code to color the picture.

● 1 blue ●● 2 brown ●●● 3 red

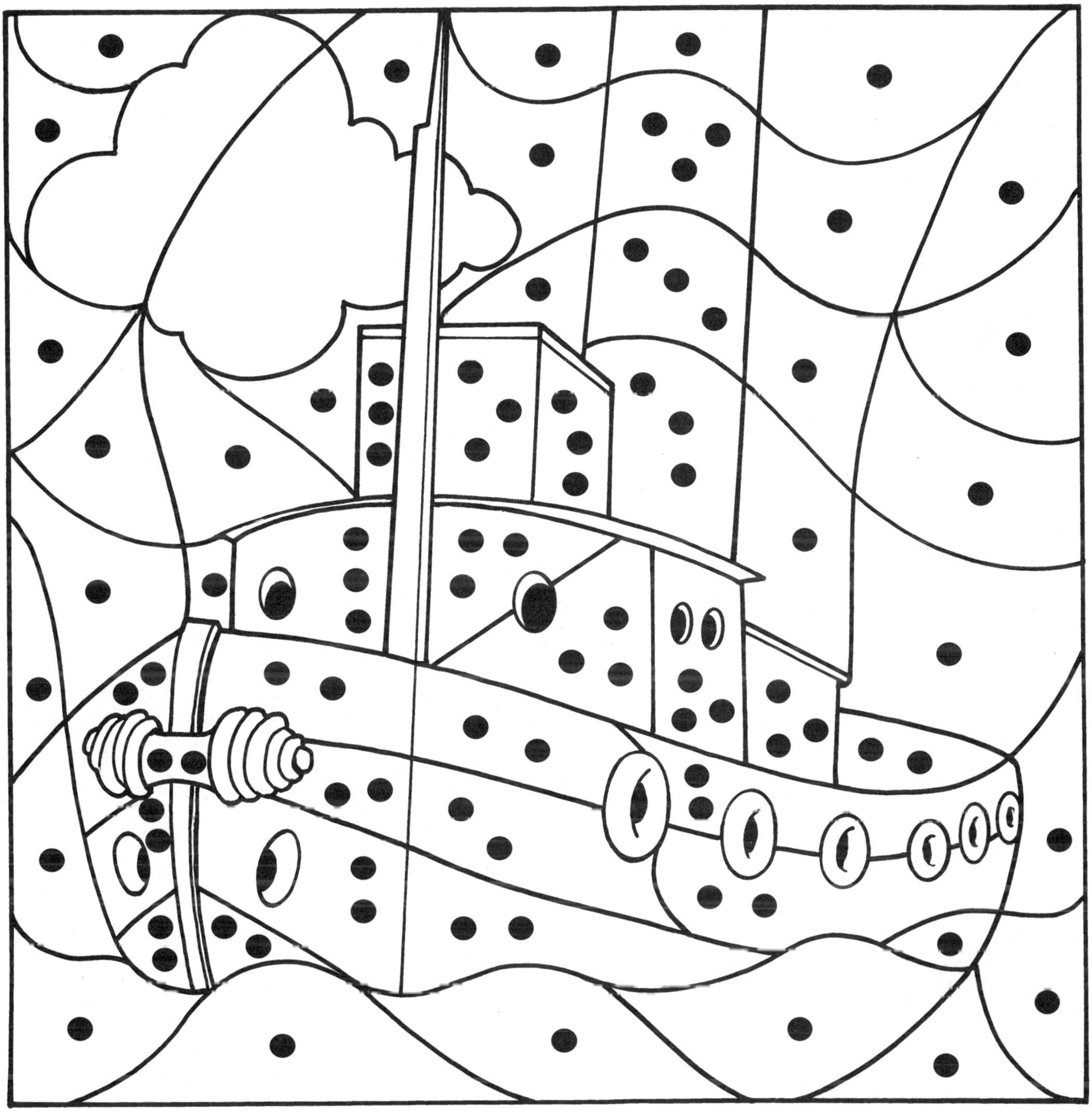

Which color did you use to color the most spaces?

Four Fine Firefighters

Trace and write.

Color each dog with 4 spots.

Climb to the Top

Count the objects on each step. Circle the matching number.

How many steps have 4 objects? ________

Name ______________________ Identifying 5

Five Friendly Frogs

Trace and write.

Color each lily pad with 5 flies.

Name ______________________________

Seashells by the Seashore

Count each kind of shell in the picture. Write the total number next to the correct shell. Circle the shells that total 7.

 ________

 ________

 ________

 ________

Circle the number that tells how many.

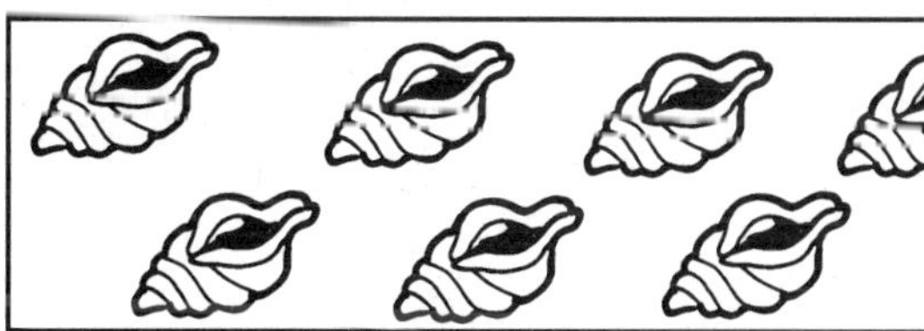 6 7

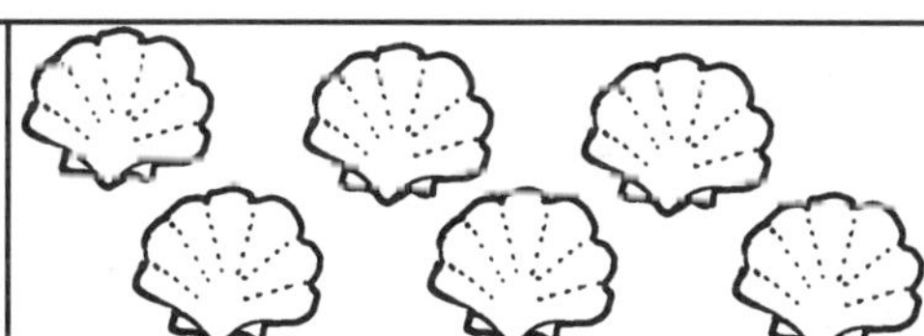 6 7

 6 7

 6 7

Name ______________________________

Eight Electric Eels

Trace and write.

8

Draw more eels to make 8.

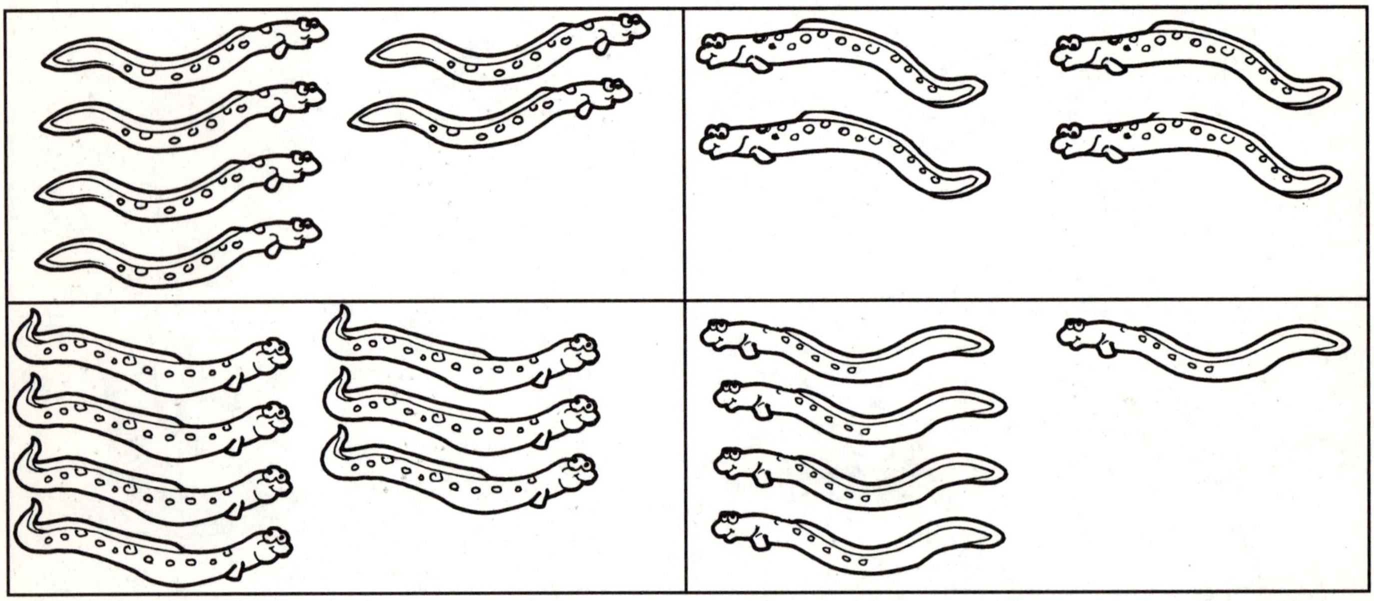

Count the eels. Color the matching number.

Eddie Eel Is Lost

Help Eddie Eel find his way back to the cave. Trace the path that goes in order from 1 to 8.

On another piece of paper, draw a picture of 8 different sea creatures.

Nine Nice Nectarines

Trace and write.

9

Color each basket that has 9 pieces of fruit.

Name ______________________________

Going to the Market

Count. Write how many. Color each fruit with 9.

💡 **On another piece of paper, draw 9 pieces of your favorite fruit.**

Name ______________________________ **Identifying 10**

Ten Railroad Ties

Trace and write.

10

Help Tina Train find the right track. Count each railroad tie.
Color the track with 10 railroad ties red.

All Aboard

Color each train car with 8 barrels red.
Color each train car with 9 barrels blue.
Color each train car with 10 barrels green.

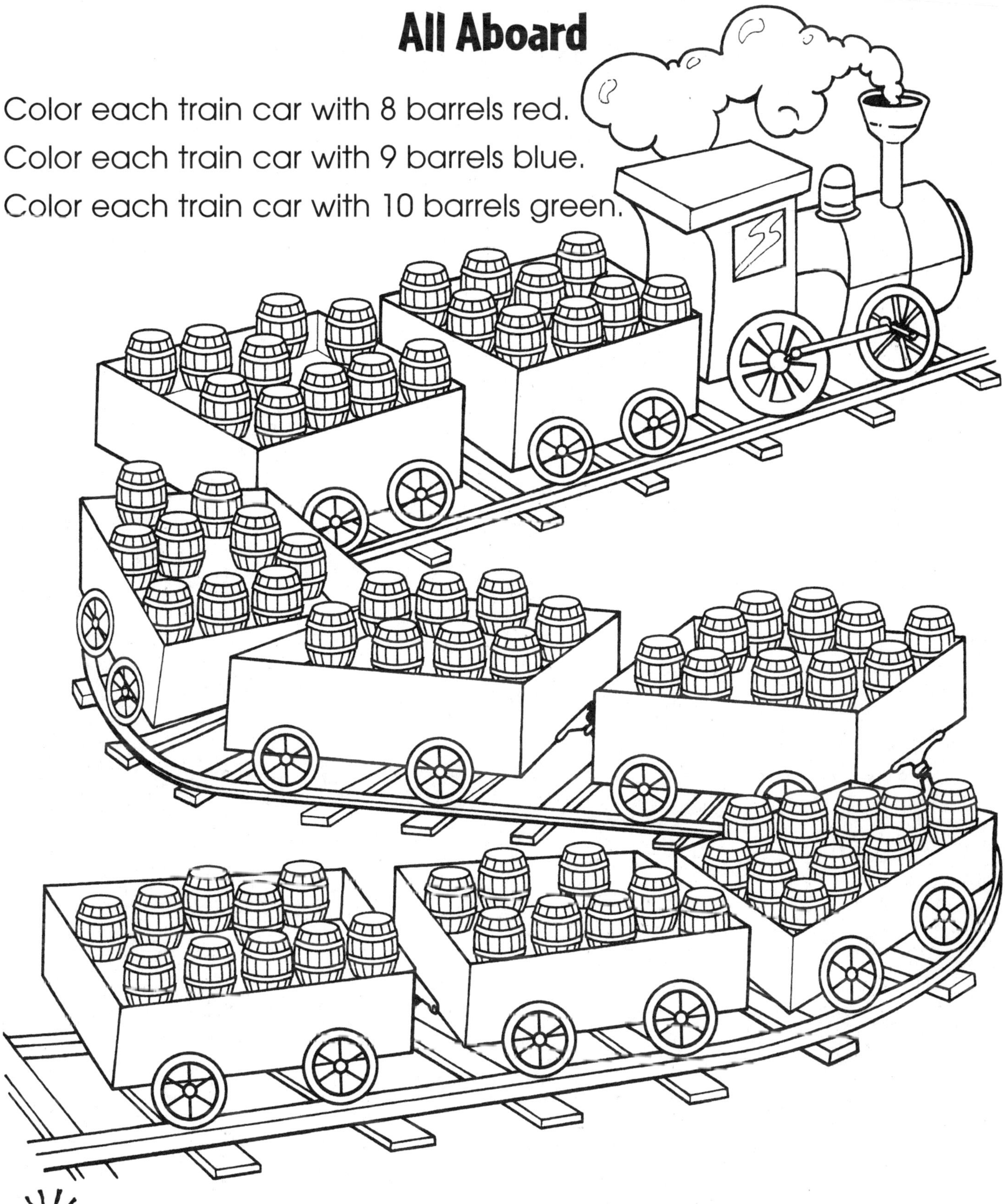

On another piece of paper, draw a train with 10 train cars.

Eleven Excited Earthworms

Trace and write.

11

Color each set of 11 earthworms.

Betty Bookworm

Count each stack of books. Draw a line to match each stack to the correct number.

Color the stack with 11 books.

Twelve Tasty Treats

Trace and write.

12

Count the candy in each jar. Color each jar with 12.

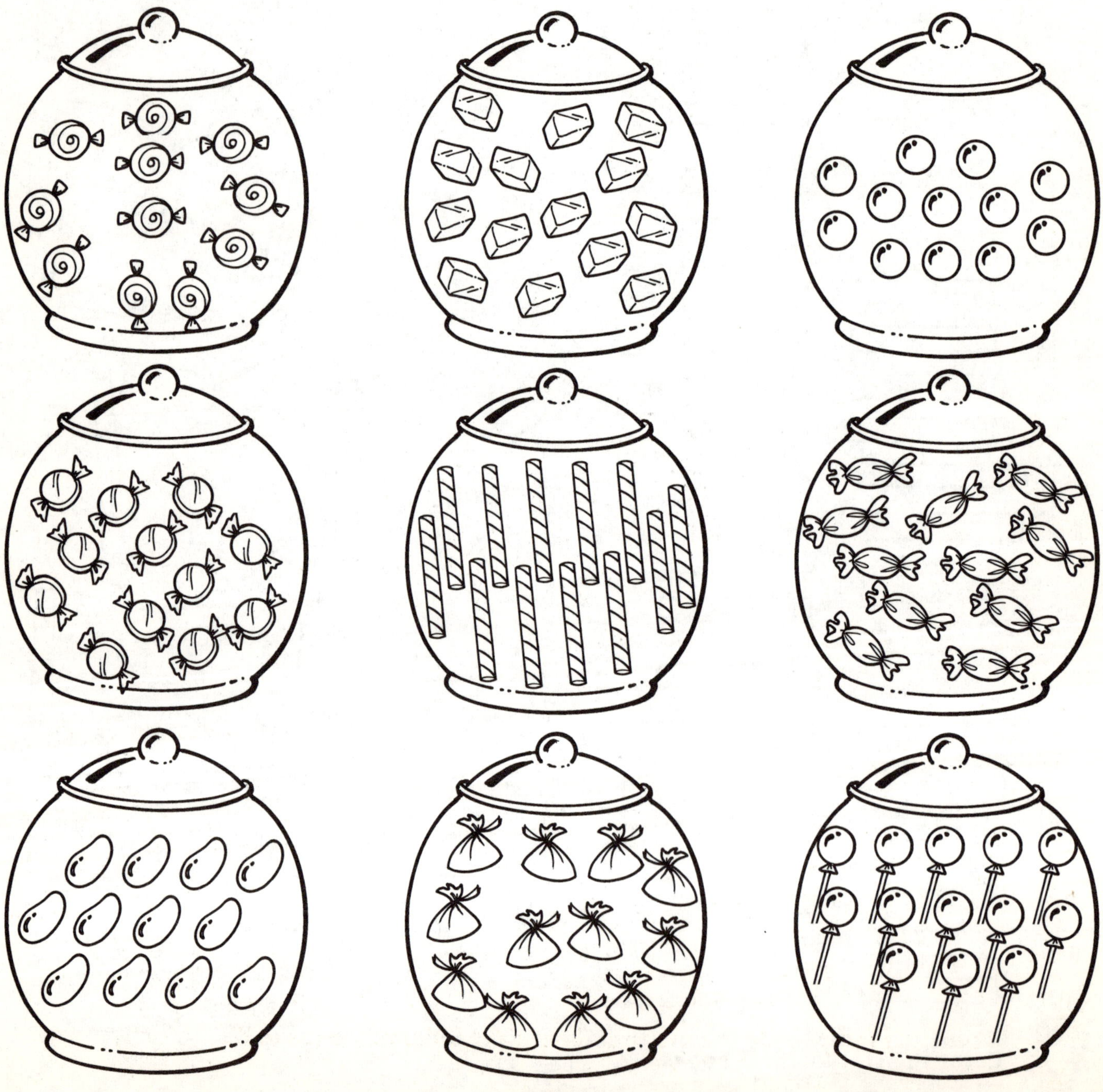

Name ____

Gingerbread Man

Help the Gingerbread Man find his gingerbread house.
Color the path that goes in order from 1 to 12.

Count the lollipops in the picture. Draw more lollipops to make 12.

Name ______________________________________

Thirteen Tasty Bones

Trace and write.

13

Circle 13 bones in each picture.

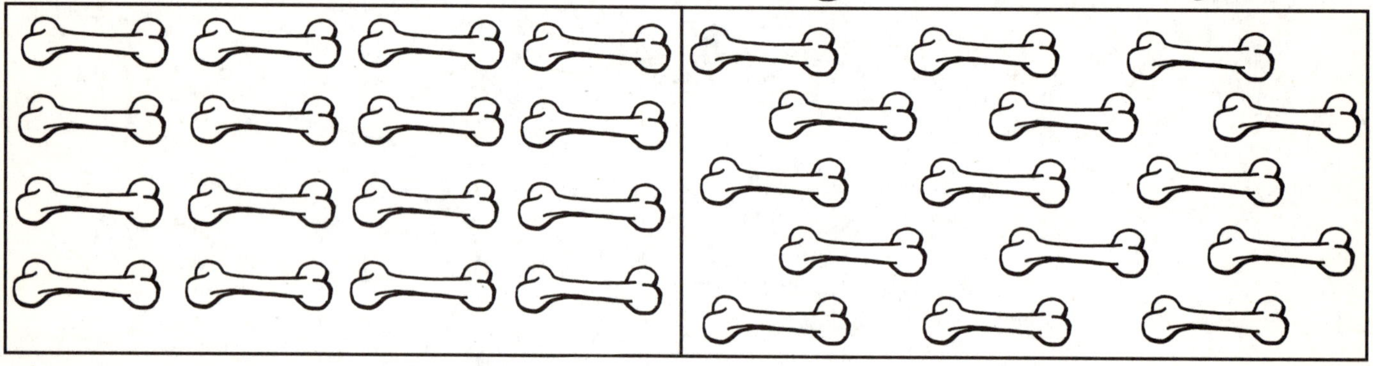

Draw more bones to make 13.

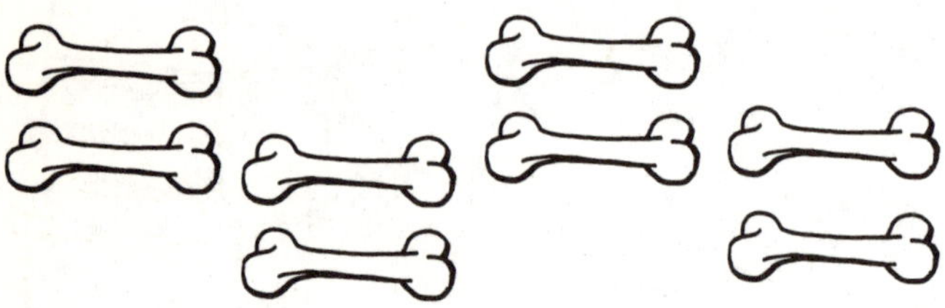

Count the bones. Circle the correct number. 12 13 14

Where, Oh Where, Has My Puppy Gone?

Help the puppy find its home. Trace the path that goes in order from 1 to 13.

Write the number that comes next in each bone.

1, 2, ______

7, 8, ______

11, 12, ______

5, 6, ______

9, 10, ______

4, 5, ______

Juggling Fourteen Balls

Trace and write.

14

Color each ball with 14 dots.

Name ______________________________

Catch the Ball!

Count. Write how many.

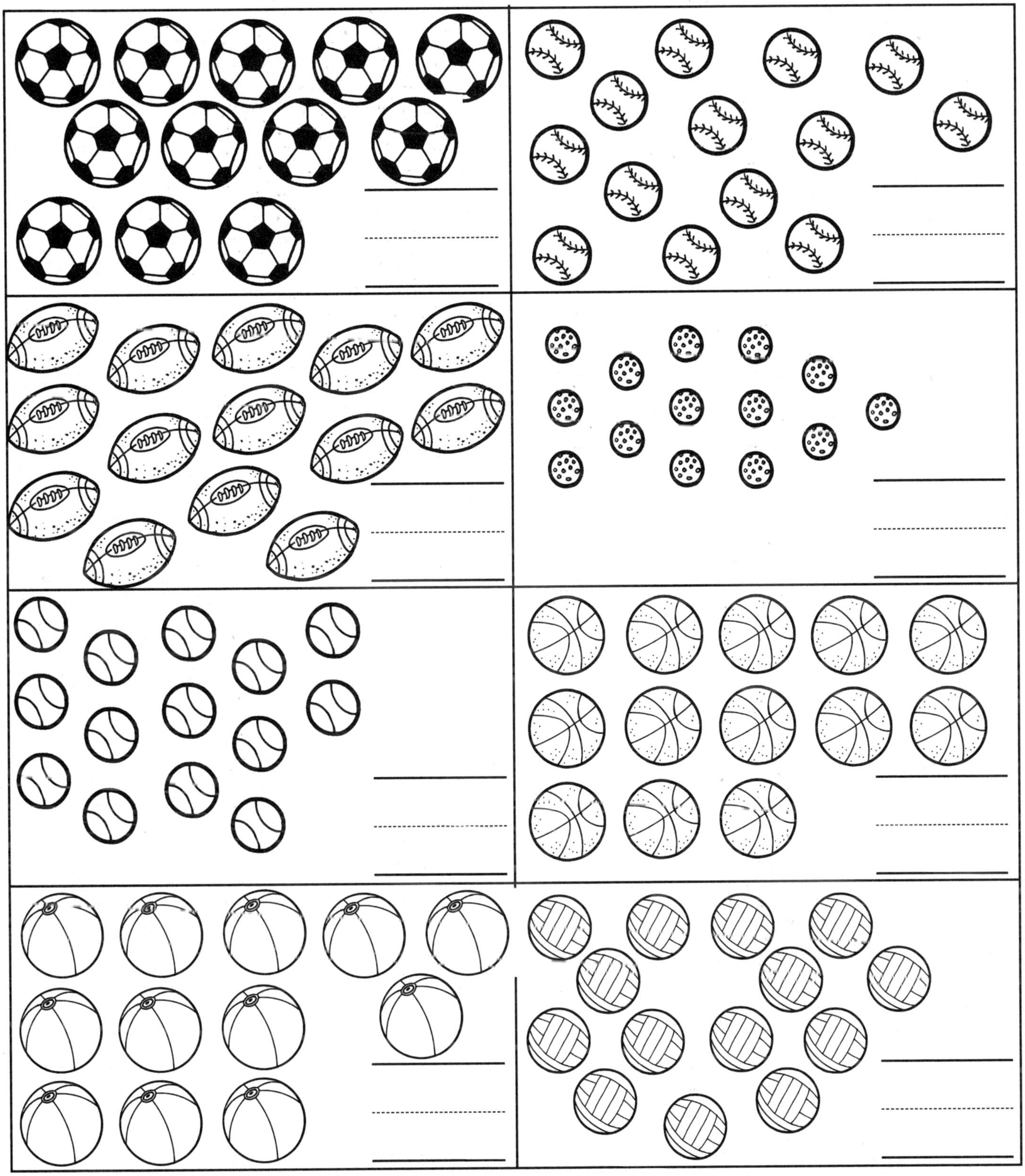

Fifteen Pennies

Trace and write.

15

Count the pennies in each bank. Color each bank with 15.

Name ______________________________

A Penny in Your Pocket

A penny equals 1¢. Count the pennies in each pocket. Write the total.

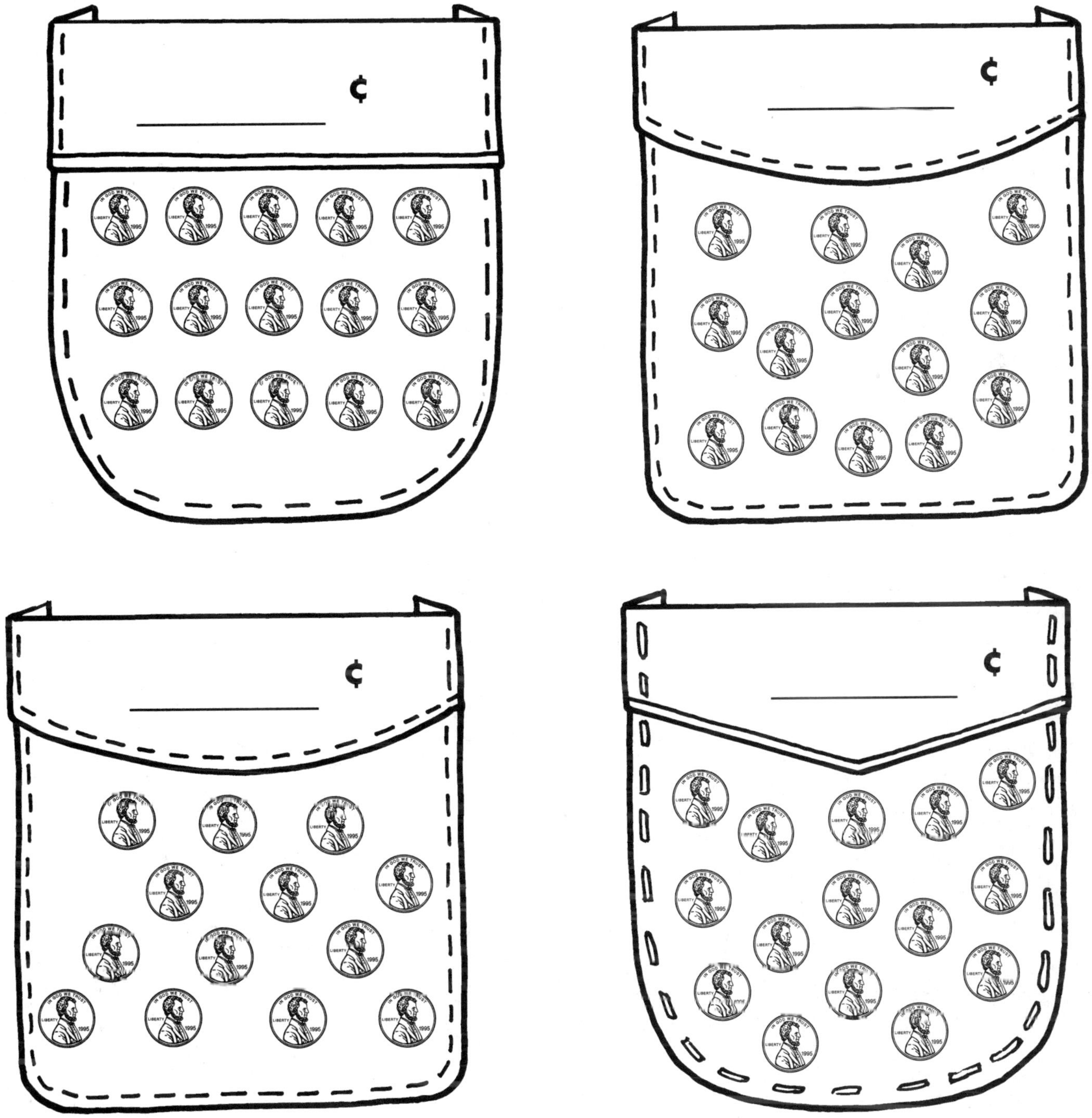

Find 15 pennies at home. Count them as you put them in your bank.

Name ______________________________________

Sixteen Kites

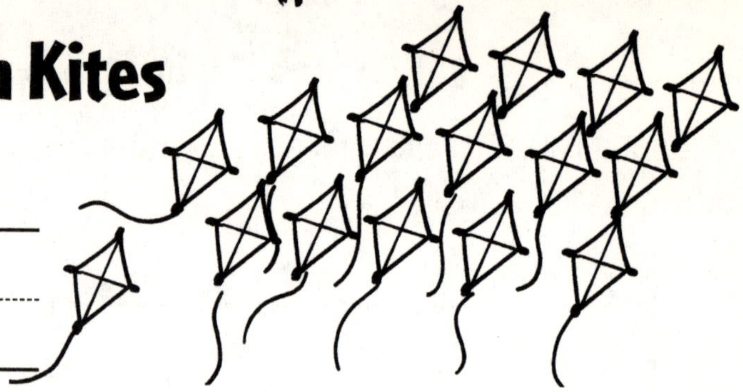

Trace and write.

16

Count the bows on each tail. Color each kite with 16 bows.

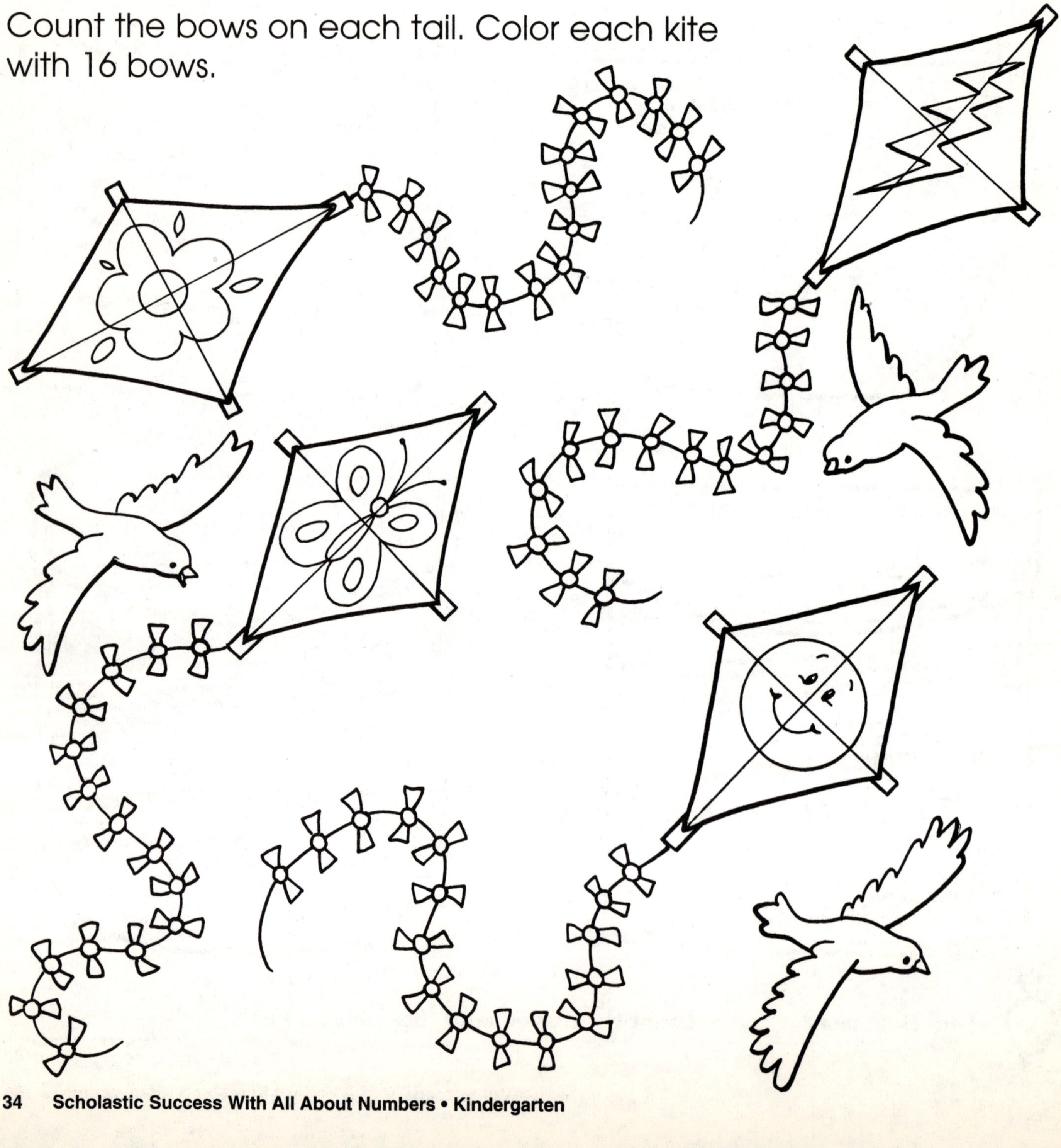

Name ______________________________

Flying High

Connect the dots from 1 to 16. Color the picture.

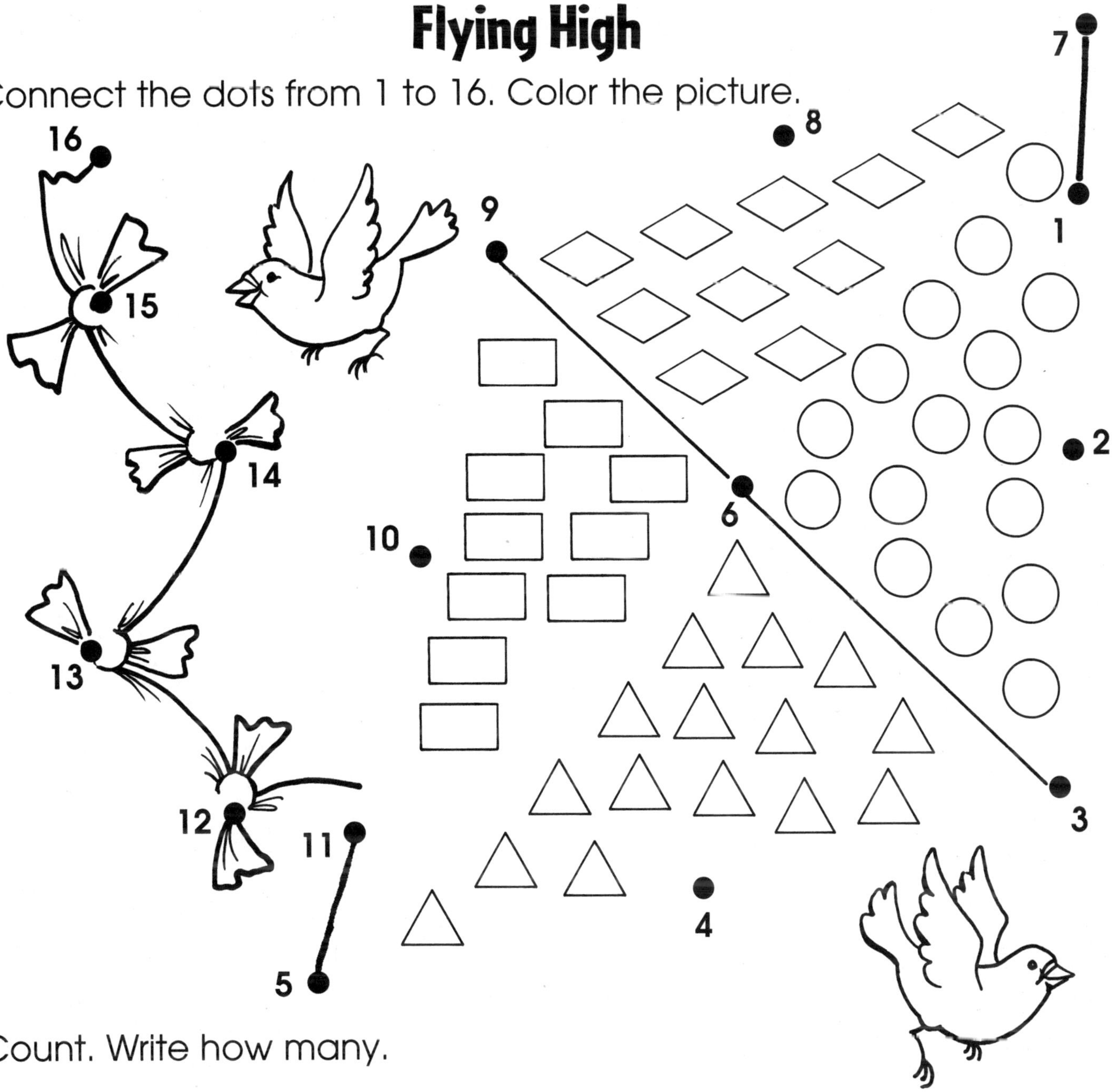

Count. Write how many.

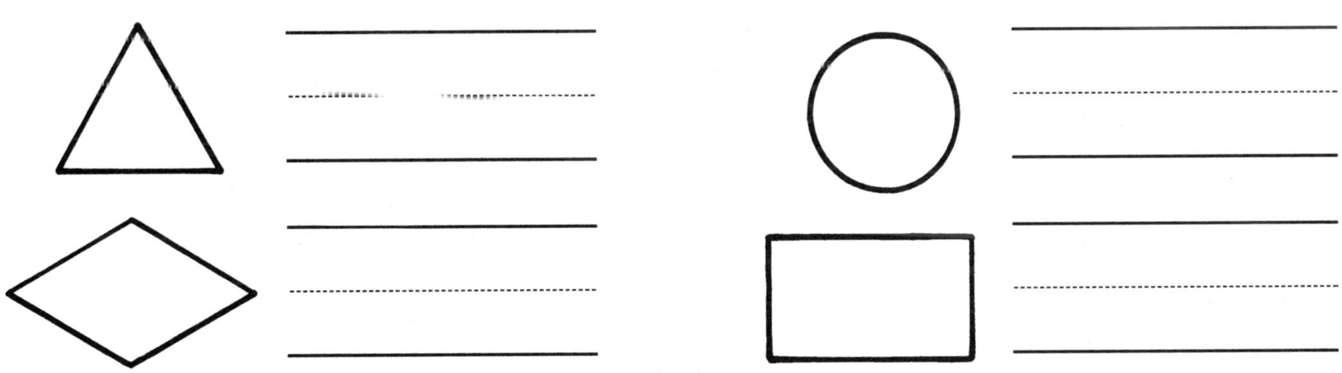

Name ______________________________

Seventeen Gallons of Gas

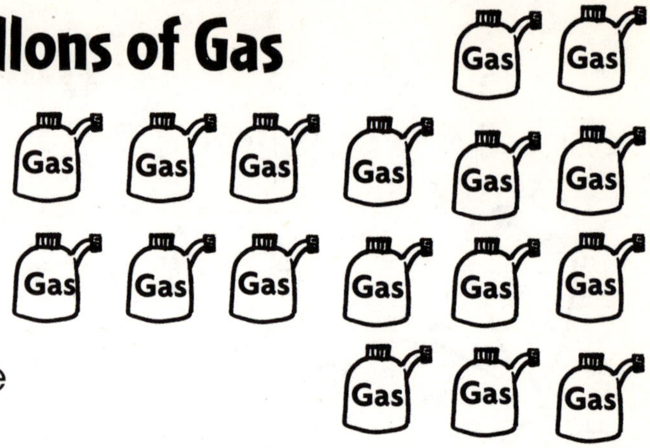

Trace and write.

17

Find the gas pump by following the numbers in order from 1 to 17.

Name ______________________________

Way to Go!

Count. Write how many.

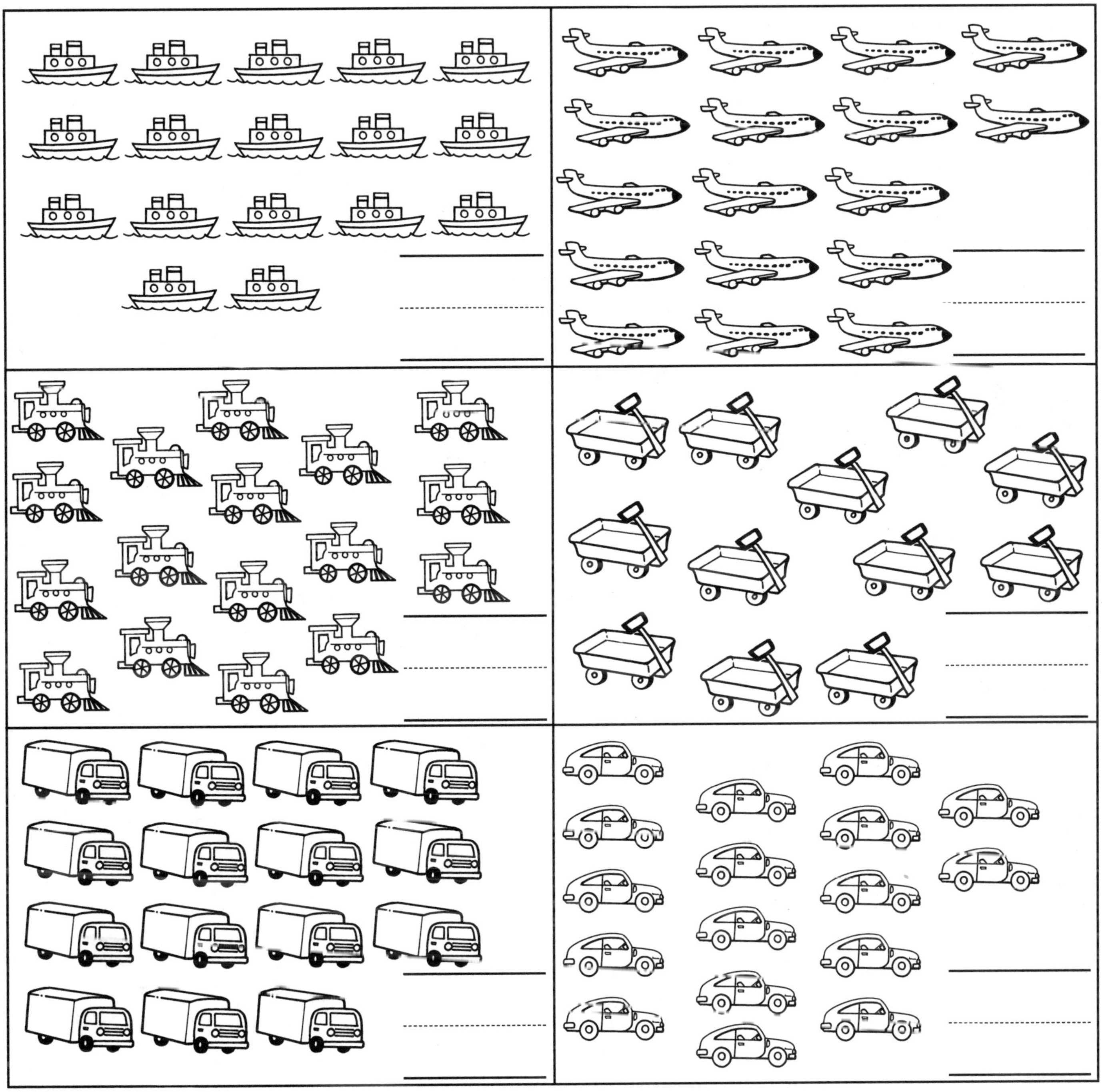

On another piece of paper, draw a train with 17 cars.

Name ____________________

Eighteen Stars

Trace and write.

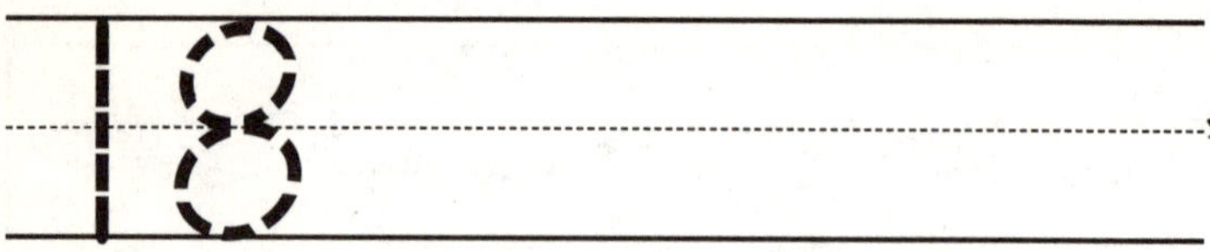

Circle 18 stars in each picture.

Draw more stars to make 18.

Count the planets. Write the number. ________

Out of This World

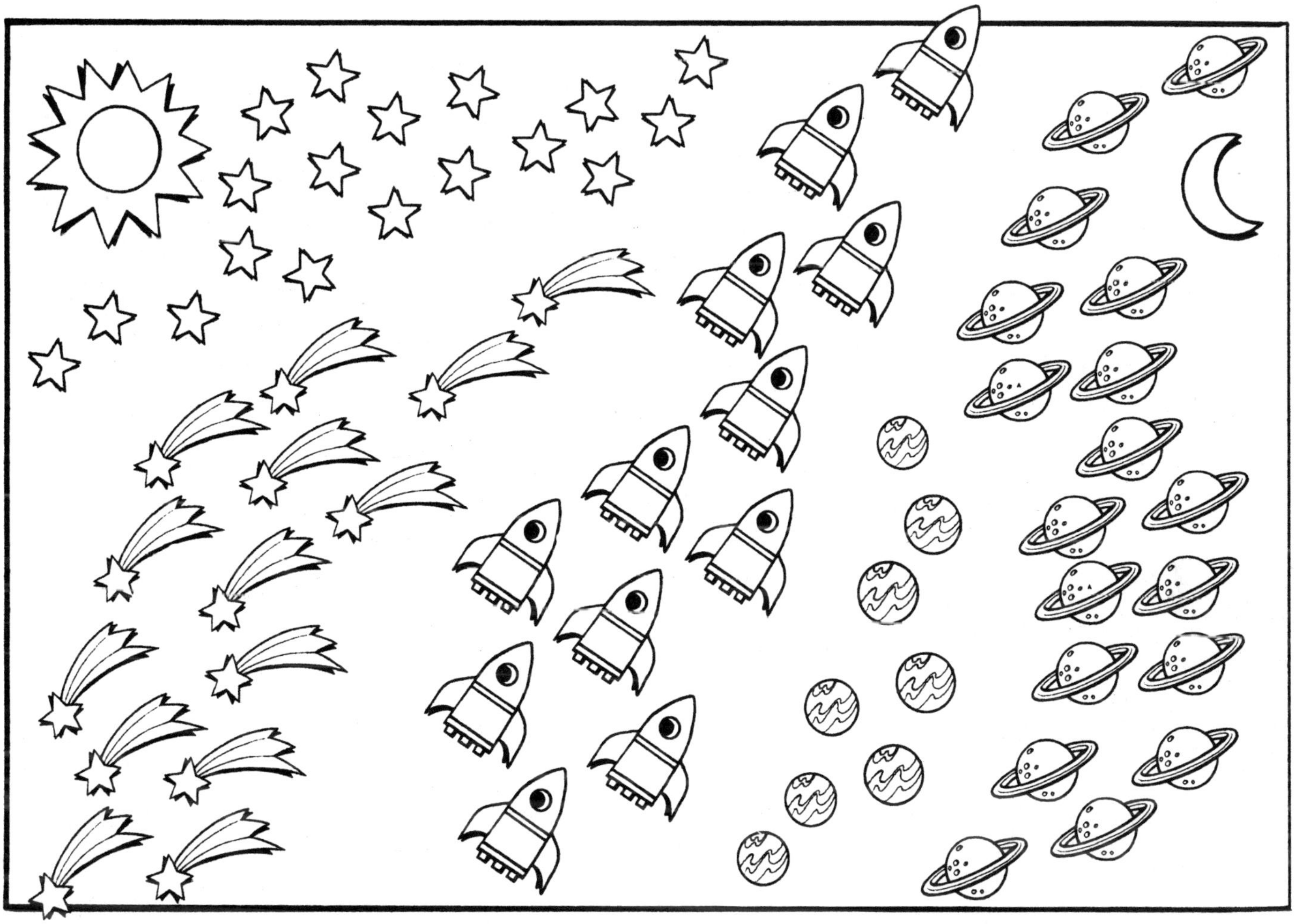

Count. Write how many. Color each group of 18 objects.

 ____________

 ____________

 ____________

 ____________

 ____________

 ____________

 ____________

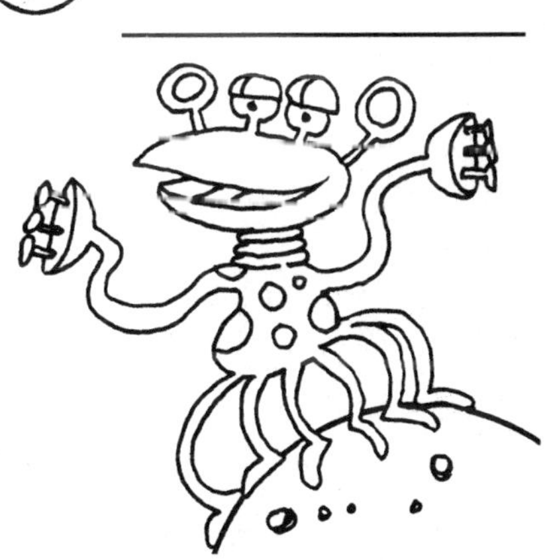

Nineteen Marbles

Trace and write.

19

Circle the number that tells how many. Color each group with 19 marbles.

17 18 19

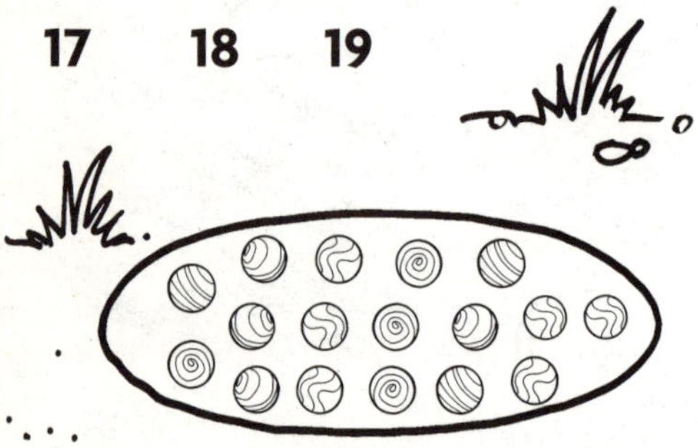

17 18 19

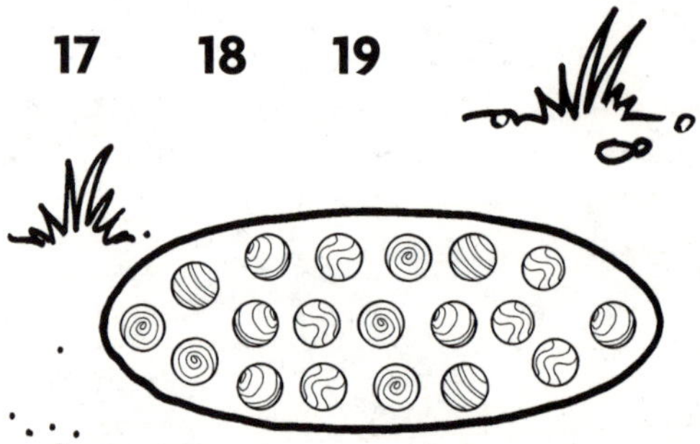

17 18 19

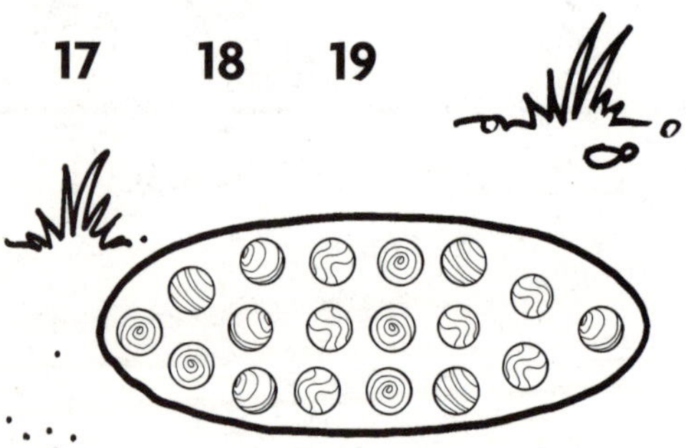

17 18 19

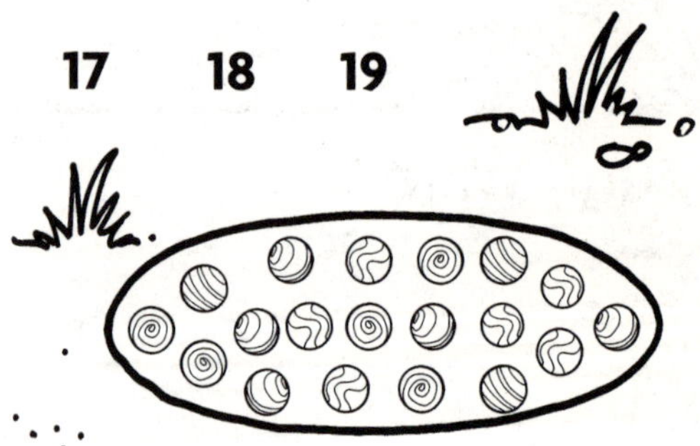

17 18 19

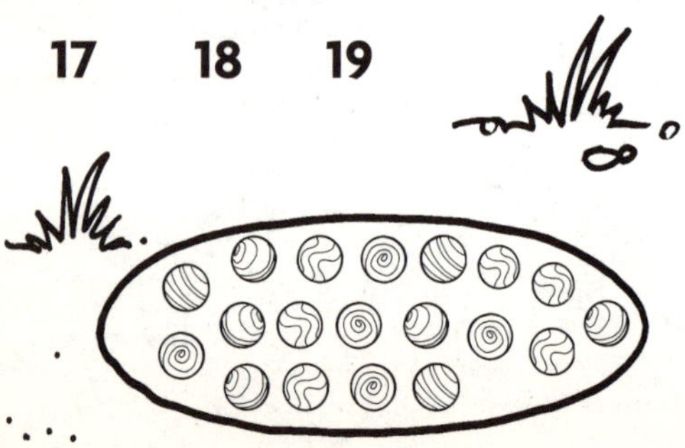

17 18 19

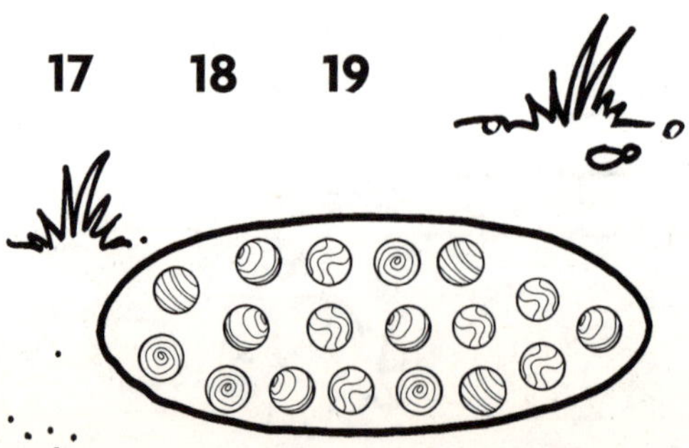

Let's Play Marbles

Circle 19 marbles.

Draw more marbles to make 19.

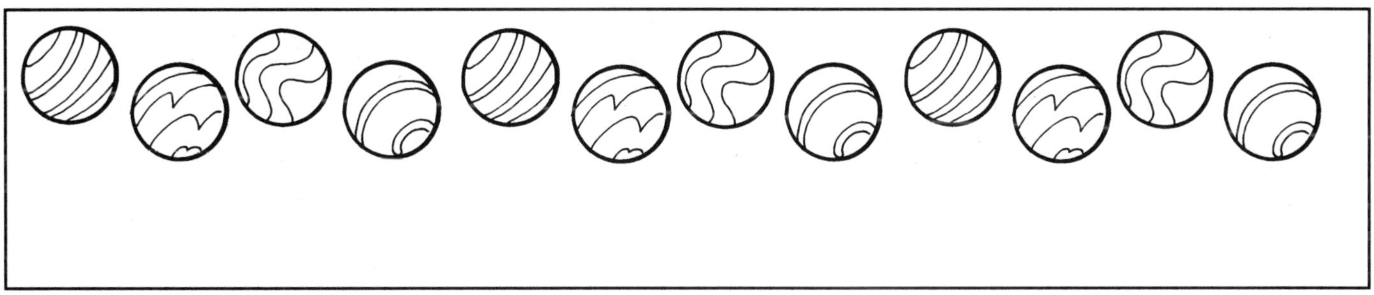

Count the marbles. Write the number.

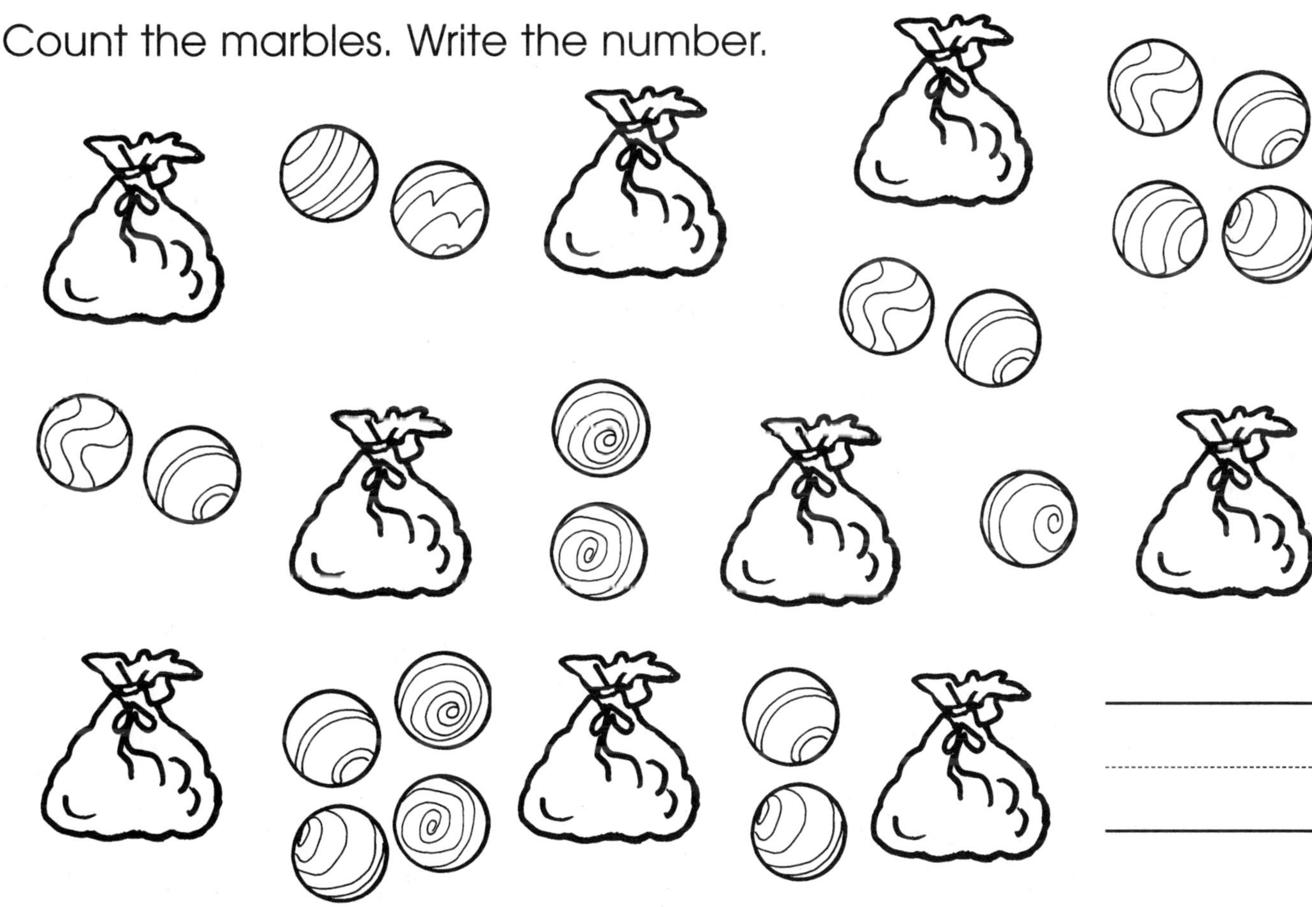

____

Name ______________________________

Twenty Butterflies to Count

Trace and write.

20

Write the numbers 1 to 20 on the trail.

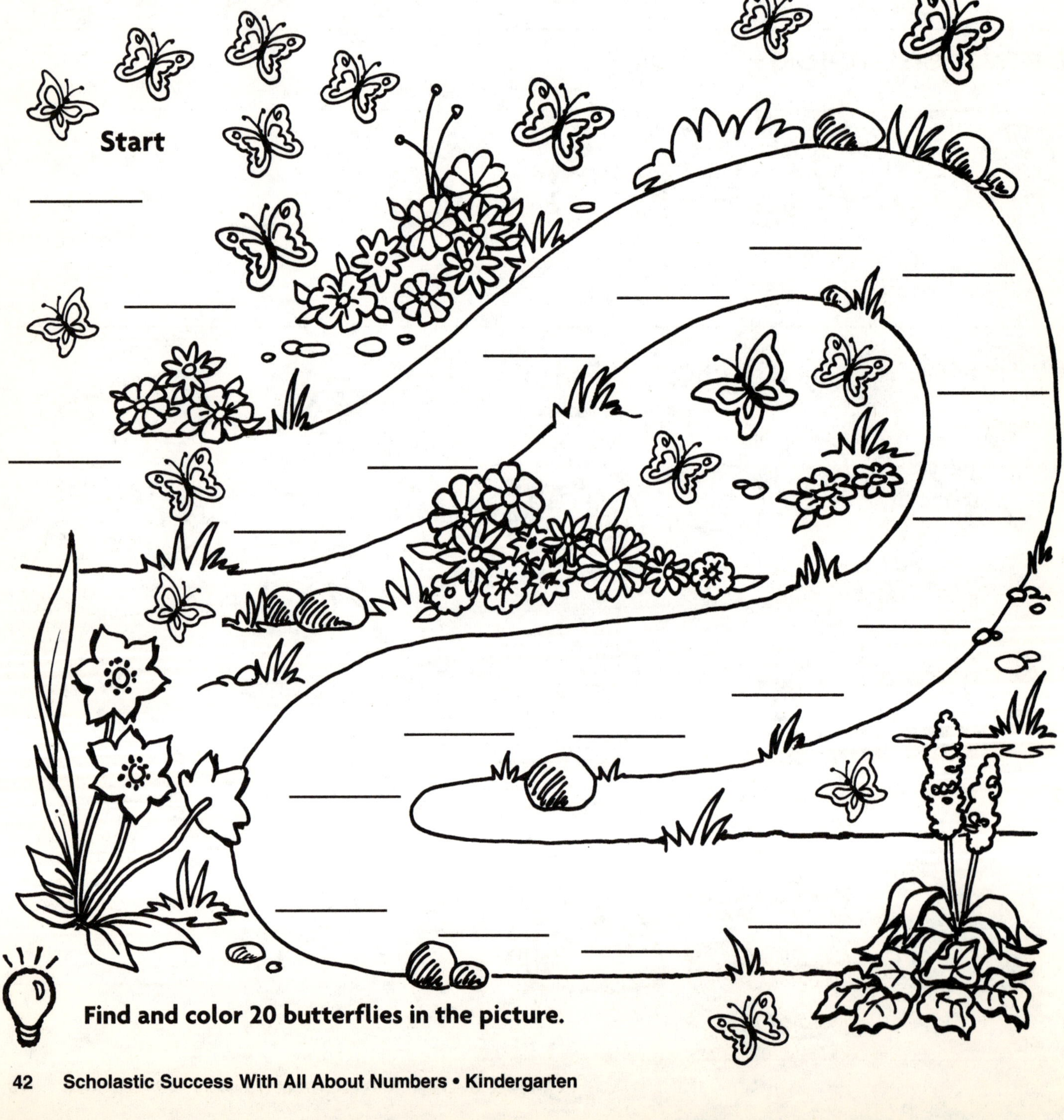

Find and color 20 butterflies in the picture.

Name ______________________________________

Don't Bug Me!

Count each group of bugs. Draw a line to the matching number.

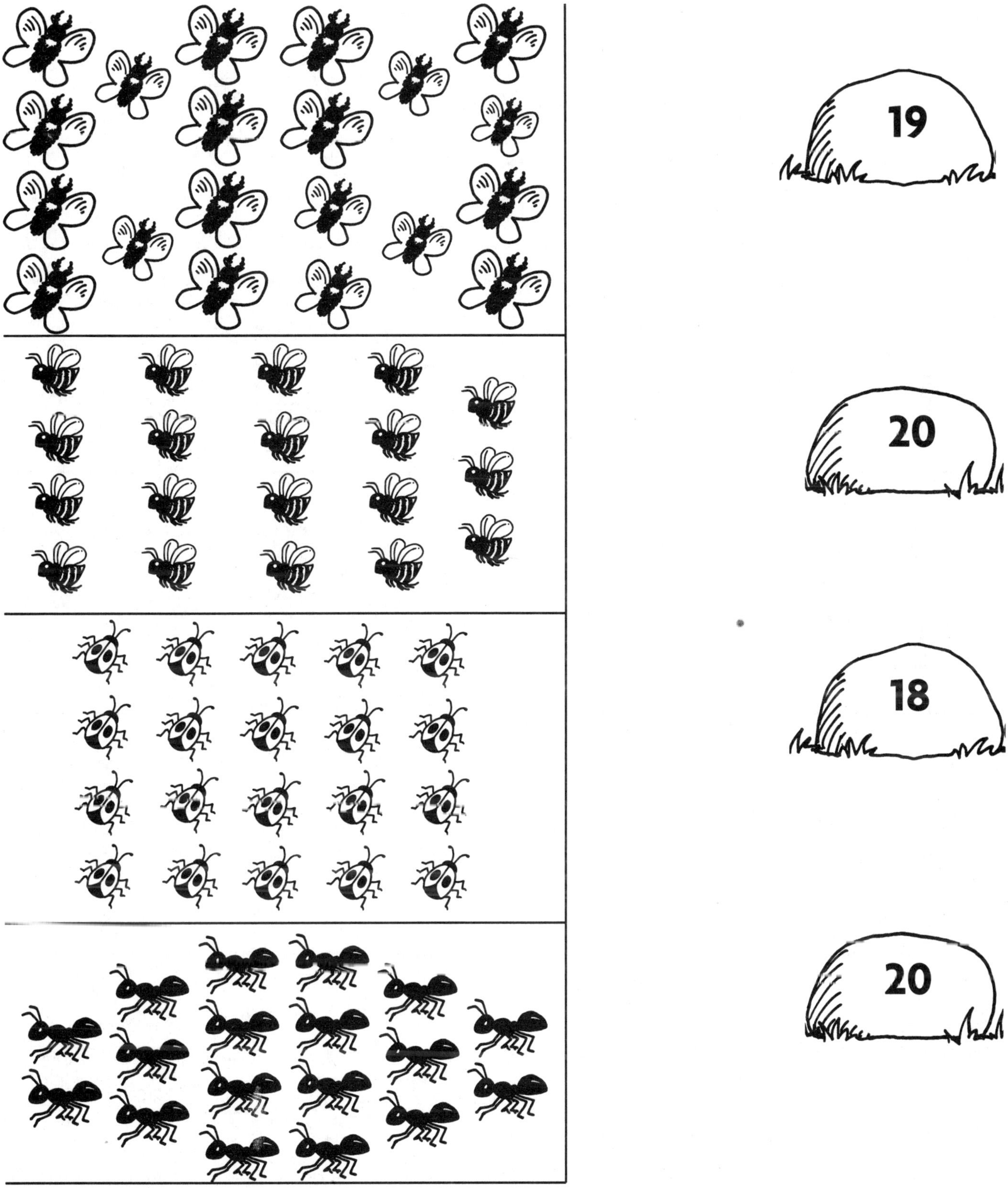

Name ______________________________

Zero at the Zoo

Trace and write.

0

Zero is the number word for 0. Zero means none.
Circle the number that tells how many.

0 1 2 3 4

0 1 2 3 4

0 1 2 3 4

0 1 2 3 4

0 1 2 3 4

0 1 2 3 4

0 1 2 3 4

0 1 2 3 4

What Am I?

Connect the dots from 1 to 30.

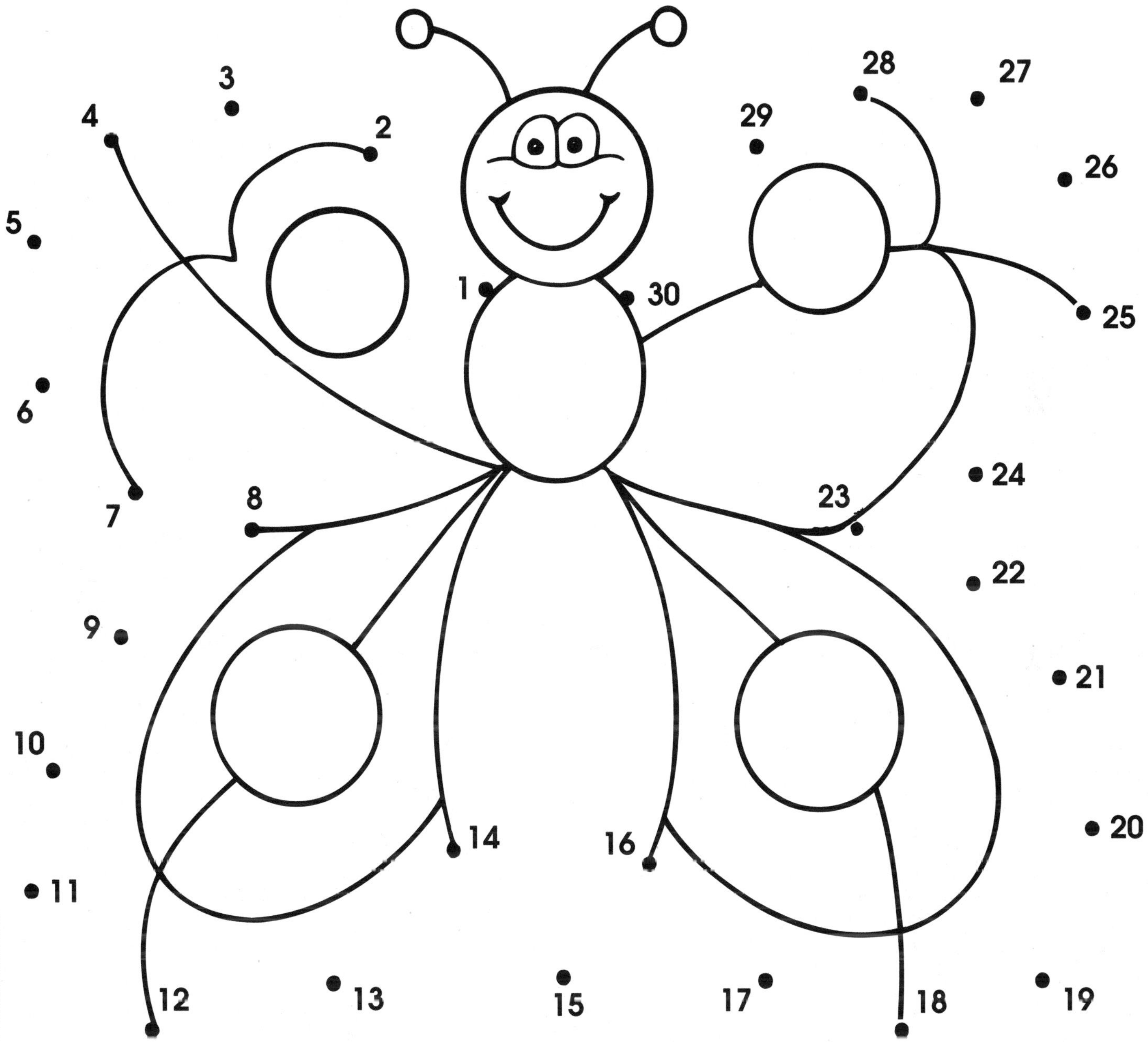

Name ________________________________

Number Words

Trace and write.

zero

one

two

three

four

five

Name ______________________________

More Number Words

Trace and write.

seven

six

seven

eight

nine

ten

On another piece of paper, have someone at home write the number words eleven to twenty. Copy them.

Name ______________________________

One-to-one correspondence

School Tools

Draw a line to match the sets that have the same number of school tools.

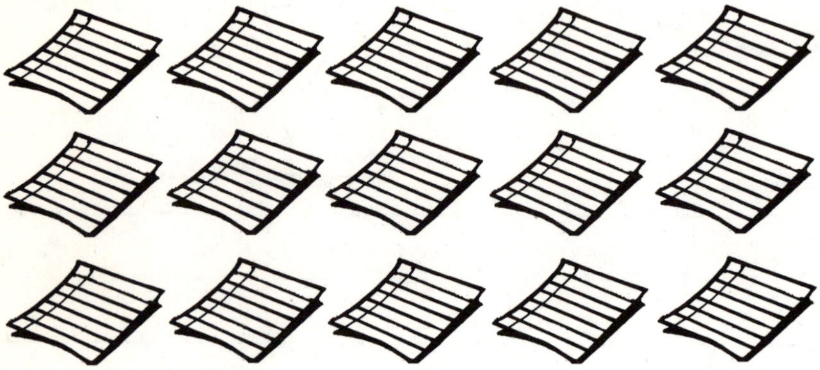

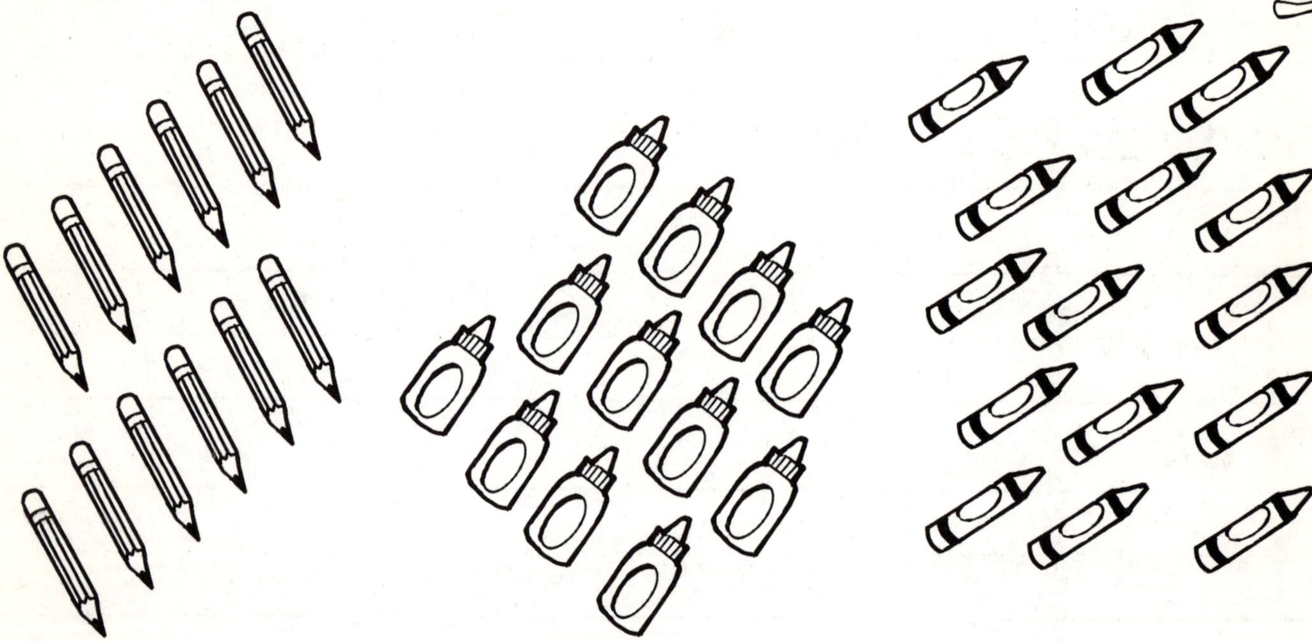

Draw one  for each  .

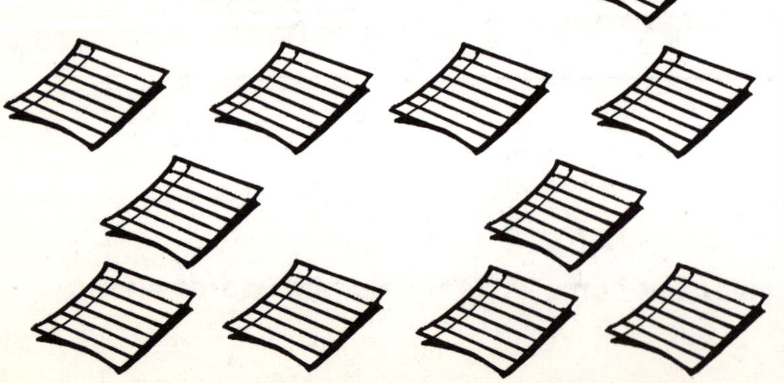